Geographic Information Systems in Government: Realizing the Opportunities

A study report for CCTA

Cambridge Computer Consultants

Dan Rickman Associates

CCTA
April 1993

LONDON: HMSO

Realizing the Opportunities of GIS

005.4 RT

© Crown Copyright 1993

Application for reproduction should be made to HMSO

First published 1993

ISBN 0 11 330607 5

For further information on CCTA's GIS publications and services please contact:

Customer Services
Information Systems Engineering Group
Gildengate House
Upper Green Lane
NORWICH
NR3 1DW

For further information regarding this publication and other CCTA products please contact:

CCTA Library
Riverwalk House
157-161 Millbank
LONDON
SW1P 4RT

071-217-3331

Contents

1 Introduction   1

    1.1   Background

    1.2   Purpose of this report

    1.3   Who should read this report?

    1.4   The structure of the report

2 The Role of GIS in Government   5

    2.1   What is GIS and why is it important to government?

    2.2   Automated Mapping

    2.3   GIS for Spatial Analysis

    2.4   Digital Cartography

    2.5   Data Collection & Interpretation

    2.6   Real-time GIS

3 GIS Organizations - Roles and Responsibilities   15

    3.1   Outline of the UK GIS Community

    3.2   Data Suppliers

    3.3   Representative Organizations

    3.4   User and Data Forums

    3.5   Professional Organizations and Special Interest Groups

4 Market and Technology Trends   31

    4.1   Market Trends

    4.2   Technology Trends

5	**Influencing Factors**		37
	5.1	Introduction	
	5.2	Triggers	
	5.3	Drivers	
	5.4	Enablers	
	5.5	Barriers	

6	**Opportunities for Working together**		53
	6.1	Data and Systems	
	6.2	Implementation of Policies	
	6.3	Strategy	
	6.4	Awareness	
	6.5	Organizational	
	6.6	Implementation Management	

7	**Conclusions and Recommendations**		69
	7.1	Conclusions	
	7.2	Recommendations for further work by CCTA	

Glossary	73
Bibliography	81
Index	83

Acknowledgements

The authors gratefully acknowledge those Departments whose staff participated in the workshop or were interviewed during the course of the study.

Chapter 1
Introduction

1 Introduction

This is the report of a study that was commissioned by CCTA, the Government Centre for Information Systems, in response to an increasing amount of GIS interest and activity within government. The study was carried out by Cambridge Computer Consultants and Dan Rickman Associates. The aim of the study was to establish what might be done by government organizations collectively, and particularly by CCTA on their behalf, to help them to realize the full benefits of GIS opportunities. Central to this, the study sought to identify barriers which obstruct the effective application of GIS, and was required to make recommendations on how these barriers might be overcome.

The study was conducted over a 2-month period running from December 1992 to February 1993 during which time approximately 30 civil servants, and others, were interviewed including:

- Prominent figures from the GIS community

- GIS champions from a cross-section of government organizations

- Representatives from coordinating and professional bodies such as AGI, LGMB, NJUG and the BCS (See Chapter 3)

- Representatives from organizations which set policies affecting GIS, including Ordnance Survey and HM Treasury.

At a mid point in the study, a workshop session was held to allow the project team to present interim findings and to discuss and extend candidate recommendations.

CCTA's responses to the study recommendations will be published separately.

1.1 Background

Government organizations are without question among the biggest users of geographic information. Every day government makes decisions concerning the infrastructure of Britain, its people, its property and its environment. Whether it is a matter of processing a routine planning application or choosing a multi-million pound road network scheme, geographic information will be a major factor in deciding the outcome. Geographic information is a fundamental aspect of government business activity. Provided government organizations are able to acquire and assimilate this information they are empowered to make informed decisions - without it they cannot.

Whereas conventional database technology has transformed our use of textual and numeric data from the paper-based arrangements of 15 years ago, it has done little to resolve the inefficiencies associated with the handling of geographic information. It remains common practice for geographic information to be recorded and distributed on paper maps, and in this form its application is limited.

Advances in technology have enabled substantial progress to be made in geographic information handling which makes concomitant improvements in services and in business efficiency possible. Whereas applications such as accounting systems are commonly implemented using conventional database technology, the advent of Geographic Information Systems (GIS) has opened up opportunities for a new range of applications based upon information that would be recorded on maps. Applications in use within central and local government today include:

- Maintaining records of coastal and river flood defences

- Analyzing changes in agricultural land use

- Modelling the impact of environmental changes

- Identifying the best location for public buildings

Chapter 1
Introduction

- Producing cartographic quality maps of demographic information

- Presenting real-time data within emergency command and control systems

- Processing planning applications

- Highways planning and management.

The GIS market is still immature and government organizations are, in general, at an early stage in exploiting its potential for improving business performance. The UK is one of the best mapped countries in the world and the government, as a principal guardian of geographic information, is one of the largest potential users and beneficiaries of this technology.

1.2 Purpose of this report

The purpose of this report is:

- To give readers an overview of the opportunities for using GIS in government, the GIS market place, the various organizations that influence the GIS market, and GIS trends

- To present the findings of the study and in particular to document the triggers, drivers and enablers of GIS usage and the barriers standing in its way

- To present the consultants' ideas and recommendations for central action, by CCTA and others, to help remove the barriers.

1.3 Who should read the report?

This report should be read by:

- Business managers in government organizations who are interested in the possibility of using GIS

- IS providers to or in government organizations who are interested in offering, or may be required to offer, GIS services

- The wider GIS community

- CCTA customers who wish to influence CCTA's future work programmes on GIS, or the work of the various other bodies described in the report.

1.4 The structure of the report

Chapter 2 describes why geographic information and GIS are important to government, and identifies generic applications of GIS.

Chapter 3 identifies the various roles and responsibilities of organizations that are represented within the GIS community.

Chapter 4 discusses the market and technology trends affecting GIS.

Chapter 5 explains the triggers, drivers and enablers of GIS, and identifies the barriers to its wider use.

Chapter 6 presents a number of ideas to overcome many of the barriers.

Chapter 7 summarizes conclusions drawn from the study findings and presents the consultants' recommendations for further work which might be undertaken by CCTA to promote appropriate use of GIS in government.

The glossary explains some of the more unfamiliar terms used throughout the report.

2 The Role of GIS in Government

2.1 What is GIS and why is it important to government?

Geographic Information Systems (GIS) are business tools specifically developed for the effective handling of spatially referenced information and, in particular, for the sharing, correlation or analysis of information of this type. Today most modern businesses are familiar with conventional database technology and its ability to organize, process and report numeric and textual information. GIS technology has many parallels with conventional database technology, but its principal distinction is that it is designed to organize, process and report information which is of a geographic nature, such as the route of a road or the location of a planning application.

In the following sections we briefly examine the generic application areas with which GIS technology is commonly associated. The majority of GIS products are applicable to more than one of the application areas, but none have facilities which apply to all types of use. In selecting GIS products, purchasers should of course take account of the precise facilities they require.

2.2 Automated Mapping

The most popular and probably the most important application of GIS is in the management of geographic records - automated mapping. It is widely established practice within government to record manually the location of important geographic features such as, for example, environmentally sensitive areas, planning application sites or street furniture. Usually such information is marked onto Ordnance Survey maps and maintained as a master record for the benefit of the whole department. Various map scales are used according to the size and required accuracy of the records involved.

Experience has shown that recording information on paper maps in this way is not always conducive to operational effectiveness. A recent study [1] showed that there are several areas in which this practice can often constrain business activity. In priority order the problems associated with records on paper maps were found to be:

- *Inefficient access arrangements*

    Map-based information is commonly distributed among many parts of organizations, and accessed by telephone or written enquiry, third party inspection and photocopy requests.

- *Currency*

    Photocopies of map records have to be updated frequently if the information they contain is to be reliable. For applications involving daily changes to map features, such as the processing of planning applications, working with copies of maps is often inappropriate.

- *Conversion and copying arrangements*

    In many cases using master copies of map-based records is unsuitable. Sometimes photocopying can be used to produce duplicate records but this in itself is often inconvenient and expensive. More importantly, paper-based records are often inappropriate for users' needs, especially when the information needs to be changed, merged, enhanced or processed.

- *Bulky storage and use*

    Maps are generally bulky to store - many government organizations have large areas of their offices devoted to map chests and cabinets, and users of maps often need large areas of work space in order to be able to work effectively. The space problem is more serious when users are required to work at map boundaries requiring two or more maps simultaneously.

Chapter 2
The Role of GIS in Government

- *Inadequate durability of original records*

    The durability of maps can also be a problem. For example, frequent users of maps may have to replace their maps annually, and will have to transfer any additional features marked on them.

- *Insufficient accuracy of use and level of detail*

    Whilst paper maps are usually just as accurate as digital ones, facilities to calculate lengths and areas, to show exact point positions and to enlarge detail, enable some GIS users to work with greater precision.

Such difficulties are most common in government bodies which have a responsibility to manage and maintain geographically dispersed assets such as the natural features, roads and rivers. It is often not realized just how fundamental these records are to these organizations' businesses. In the case of the National River Authority (NRA), it was found that an estimated 25% of its 8000 staff are regular map users and of these over 1000 staff use maps for more than 6 hours per week. These statistics are not so surprising when one considers that an estimated 80% of its information is accessed by its geographic reference.

This level of dependency on map-based records is a feature of many other organizations such as:

- Department of the Environment

- Department of Transport

- English Nature

- Forestry Commission

- HM Land Registry

- Local Authorities

- Ministry of Agriculture, Fisheries and Foods

- Ministry of Defence
- Office of Population, Censuses and Surveys

These are only a few examples of public bodies in which map-based records are a vital business asset; many others have a similar level of dependency and it is within such organizations that GIS technology can often be applied to best effect.

It is often not appreciated that when GIS is used for this type of records management, the *base* or *background map* usually serves only to provide geographic context to *foreground features*, and frequently the foreground information can be recorded at medium or small scales. Consequently projects of this type can be implemented using relatively inexpensive data products.

Just as conventional database technology has replaced many instances of card indexes and paper files, GIS technology can often transform the efficiency of geographic record keeping.

2.3 GIS for Spatial Analysis

The ability to collect, manage and retrieve geographic information is a valuable facility in itself, particularly if the quantities of data are very large. However, the full potential of GIS can be realized only when it is used for combining and analyzing geographic datasets.

Whereas textual and numeric data is processed using textual and numeric operands such as *add, subtract, greater than* and *less than*, GIS introduces spatial operands for processing data, such as *within, connected to*, and *adjacent to*. Spatial operands can be combined into sequences of instructions to produce many new applications and to solve spatial problems. Such applications can automate aspects of planning application approval or produce new features such as a bus route, or a catchment for a school. Other examples include:

- Finding the best position for a broadcasting mast
- Defining Nitrate Sensitive Areas

Chapter 2
The Role of GIS in Government

- Predicting the spread of air and water pollution

- Producing rainfall contours from point source data

- Mapping land-use change

- Transport network capacity planning.

As with the most modern of database management systems, 4th generation tools are often used for designing screen, report and plot layouts, and users interact with the systems using Windows, Icons, Menus and Pointers (WIMP interfaces).

The impact of GIS is typically that staff are more productive because they can process information more quickly, and spend less time simply trying to access the information they need. Another benefit is that where GIS delivers a tangible improvement in the quality of management information, the resources deployed are often better targeted. The potential impact for business applications which are resource-intensive and geographically-based can be substantial.

It should be noted, however, that ability to perform spatial analysis is at present hampered by difficulties of integrating statistical analysis and modelling software into a GIS environment. A number of research projects are working on solutions to this problem.

2.4 Digital Cartography

Many government organizations have requirements to produce new paper maps for:

- Management publications

- Operational needs

- Public information

- Provision of information to other parts of government.

Traditional map making methods are costly and take a long time; hence a published map is almost always out of date. Once geographic data is available within a GIS, bespoke maps can be produced easily and quickly, either as one-off exercises to satisfy individual public requests for information, or as master copies for reproduction in large quantities.

GIS products which are used for records management can produce presentation-quality maps, and for most purposes this standard is adequate. However, true cartographic-quality maps can be produced using specialized software which has facilities for dealing with more sophisticated graphics and publication-quality output devices.

2.5 Data Collection and Interpretation

A considerable amount of government business activity concerns the collection of geographic data. Sometimes data collection is required for one-off situations such as a specific study or incident; at other times it can be part of a routine occurrence, for example as part of a monitoring programme. Some forms of GIS support the assimilation of such data by providing facilities for:

- Land surveys

- Demographic/market surveys

- Aerial photography

- Satellite imagery.

Land surveys
Electronic surveying tools and Global Positioning Systems (GPS) have been available for several years and they have been developed to interface to CAD systems and specialist surveying software. Interfaces have been added to many GIS systems to provide two-way communications between CAD and surveying software thus allowing geographic features such as new roads or land drainage to be acquired from design tools electronically.

Chapter 2
The Role of GIS in Government

The recent emergence of hand-held GIS products on pen-based computers is a significant development for field-work of this type.

Demographic / Market surveys
GIS can also be applied in the collection and interpretation of population-related information. Software and data are now available to allow property addresses and postcodes to be converted into map references. This kind of facility allows information which is referenced by address or postcode to be analyzed and presented.

Aerial photography
Several GIS vendors offer products which have facilities for combining aerial photograph images with other geographic features. Aerial photographs are a particularly rich source of geographic information, but usually require manual interpretation to be of most value. Many local authorities have large holdings of aerial photographs which remain under-utilized because of the problems of access. Whilst current computer technology cannot reproduce the very fine quality of paper photograph prints, it can be used to make photographic archives more widely available. Many types of object can be easily identified directly from a computer screen allowing features such as woodland and vegetation to be recorded easily. Quite often the only practical alternatives for acquiring such information are labour-intensive ground surveying techniques.

Using aerial photography as a source of geographic information does not always involve the expense of specially commissioned aerial surveys; copies of aerial photographs can be purchased from Ordnance Survey and the Royal Air Force. One private sector company intends to conduct 5-yearly aerial surveys of the whole of the UK to establish its own colour aerial photograph archive which will be commercially available.

Satellite Imagery

Along similar lines, satellite imagery is a practical source of geographic information. Recorded at a much smaller scale, the electronic images produced by satellites can be used to derive new geographic information such as land-cover on a national scale. Since satellite images are frequently updated, the information can be particularly useful to monitor changes in the environment such as the growth of cereal crops, the spread of an oil spillage, or the weather. As with aerial photography, it is possible to record important geographic features directly from satellite images. The interpretation of satellite images is highly specialized; several organizations provide this interpretation as part of their service.

2.6 Real-time GIS

The examples of GIS discussed above use geographic information which has been captured at discrete moments in time, but there are some types of application which demand a continuous stream of real-time information. Real-time geographic information is usually associated with monitoring systems and command and control systems. Examples of these include:

- Monitoring of river levels/flows
- Traffic control - road, rail, air and sea
- Ambulance dispatch
- Pollution alert
- Command and control of defence operations.

There are also several examples of GIS being used as part of Command and Control systems within fire, police and ambulance services. In several cases these systems use technology developed for the security industry to track vehicles as they move around. The position of vehicles can be displayed on a projected map display allowing operators in the control-room to dispatch the nearest appropriate vehicle to an incident.

Chapter 2
The Role of GIS in Government

Another application for real-time GIS is in-car navigation systems which can be used to reduce journey times and direct traffic around traffic jams. In Japan some 200,000 production vehicles have been equipped with simple navigation facilities, and in Europe a number of private sector companies are working with the European Community to develop a pan-European car navigation system.

3 GIS Organizations - Roles and Responsibilities

3.1 Outline of the UK GIS Community

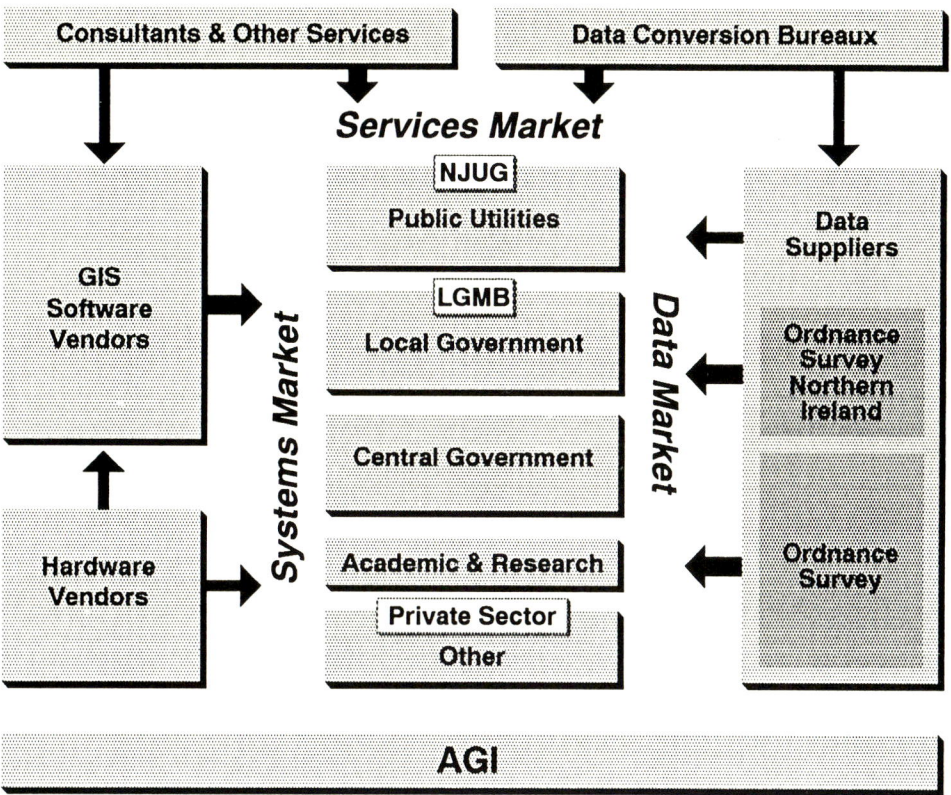

Figure 3.1 - The UK GIS Community

As shown in Figure 3.1, the GIS community is divided into four main areas:

- Systems suppliers
- Data suppliers
- Service providers
- Users

It is common for organizations to have a strong market profile in one of these areas and, in addition to have the capability of offering services or products from the other areas, eg. "turnkey" solutions involving the supply of hardware, software, data and implementation-management services.

Systems Suppliers
The supply of Geographic Information Systems is dominated by a small number of international players, but some small companies have survived by establishing joint marketing agreements with large hardware vendors, and are prospering.

Data Suppliers
For historical reasons, the large-scale map data supply market is dominated by Ordnance Survey in Great Britain and Ordnance Survey (NI) in Northern Ireland. Other organizations are able to collect and supply large-scale map data, however; OS and OSNI face competition in the more commercially-rewarding urban areas. In common with other copyright owners, OS and OSNI are protected by the 1988 Copyright Act to safeguard their investment, although policy is to maximize the use of their data. Other data suppliers have been able to compete successfully with the two Ordnance Surveys with specialized and small scale products.

Service Providers
A number of organizations offer specialist services such as:

- Implementation services
- Data conversion
- Management services
- Market research
- Training.

Chapter 3
GIS Organizations - Roles and Responsibilities

User Community
The user community largely consists of organizations which provide services to the public in some form or other. There are five main groups involved:

- Public utilities

- Local government

- Central government

- Academic and research

- Others - such as market research companies and retail chains.

Representative organizations and groups
The GIS community also contains a number of groups that represent various industry sectors and professional bodies. Organisations such as the Local Government Management Board (LGMB), and National Joint Utilities Group (NJUG), coordinate activities in their respective market sectors, and the British Computer Society and British Cartographic Society support GIS Special Interest Groups.

However the primary forum for GIS in the UK is the Association for Geographic Information (AGI) which was formed in March 1988 and has a membership of 600 from all sections of the GIS community.

In the following sections we identify and discuss a range of organizations which play a variety of important roles within the GIS community. The many organizations involved in GIS research are not, however, covered in this report.

3.2 Data Suppliers

3.2.1 Ordnance Survey of Great Britain (OS)

Ordnance Survey (OS), the national mapping agency of Great Britain, is an independent government department which has been an executive agency since May 1990. It employs 2300 staff at headquarters in Southampton and at some one hundred field offices around the country. OS is responsible for the official survey or topographic mapping of Great Britain. Among the products and services which OS offers are:

- Paper maps of the whole of Great Britain at scales ranging from 1:1250 to 1:625,000+

- Maps on microfiche

- Bespoke maps - enlargements, reductions and special editions

- Surveying services

- Map data products

- Aerial photography

- Books

- Gazetteers

- Overseas mapping.

Ordnance Survey has a particularly important role to play for the GIS community - it has a virtual monopoly in the production of medium and large-scale maps and, consequently, is in a unique position to offer digital versions of these.

Ordnance Survey's current portfolio of digital map products includes:

- 1:1250 , 1:2500 , 1:10000 scale vector - Land Line data vector maps

- Road Centre Line Network (OSCAR)

Chapter 3
GIS Organizations - Roles and Responsibilities

- Administrative Area Polygons (Boundary Line)

- National Address Database (Address Point)

- 1:250,000 scale vector maps

- 1:625,000 scale vector maps

- 1:10,000 scale scanned maps

- 1:50,000 scale scanned maps

- Gazetteers - derived from 1:50K, 250K and 625K maps.

- 1:10,000 and 1:50,000 contours and Digital Terrain Models.

OS has concentrated its efforts on completing the large scale vector map products, which are particularly important to local authorities and utilities. This has required a substantial digitising programme which is now at an advanced stage and is due for completion in 1995. Since OS is required to recover its costs through product and service charges, the charges for these maps necessarily includes the costs of the digitizing programme.

The recent release of 1:10000 and 1:50000 scale scanned maps is a potentially important development for a substantial number of GIS applications. Many government organizations record additional features on to paper maps at these scales, and the availability of equivalent electronic maps enables GIS users to work with familiar map details and to produce identical copies of original records.

Furthermore, the relative simplicity and economy of scanning has meant that scanned maps are very much less expensive than vector equivalents.

The availability of these maps at an affordable price is expected to stimulate the use of GIS technology for applications which simply use maps to add context to other data.

3.2.2 Ordnance Survey of Northern Ireland (OSNI)

Ordnance Survey of Northern Ireland (OSNI) in existence since 1922 has been established as an Executive Agency within the Department of the Environment for Northern Ireland since April 1992. OSNI employs just over 200 staff, the majority of whom are located in Belfast Headquarters which houses the cartographic, reprographic and computer services, marketing and administrative support. At present some 80 staff working from Regional Offices throughout Northern Ireland keep the cartographic records up-to-date.

OSNI exists to provide comprehensive and up-to-date topographical information for public and private sector organizations through:

- The production of maps, street plans and atlases at various scales and of gazetteers to meet the needs of the public and private sectors and private individuals. Approximately 16,000 different maps are published from the topographical archive, increasingly in digital form, at scales from 1:250000 to 1:1250

- The creation and maintenance of an archive of aerial photographs of Northern Ireland to support the mapping programme, and the operation of the Northern Ireland Remote Sensing Centre with its archive of satellite imagery for the province

- Leading the implementation of the Northern Ireland Geographic Information System (NIGIS) - a multi-partner GIS strategy which aims to link many public service organizations within the province.

Chapter 3
GIS Organizations - Roles and Responsibilities

OSNI's continuously maintained general purpose Computer Mapping and Topographical Database (COMTOD) provides OSNI with an up-to-date structured digital archive of topographic features. This is underpinning NIGIS (see 3.4.4), and has helped establish OSNI as an international leader in the application of geographic information and GIS. With links to associated textual data, particularly postal addresses, personalized map extracts are available centred on a customer designated location.

3.2.3 Military Survey

Military Survey is a Defence Support Agency within the Ministry of Defence. It has responsibility for the provision of geographic data in all its forms to support the current and future needs of the UK armed forces.

Military Survey intends to satisfy these requirements with output from a Defence Geographic Database (DGDB) which will support the generation of a number of products. Several digital map products are being specified and agreed by international working parties and are also intended to address civilian requirements. In May 1992 Military Survey announced the availability of the Digital Chart of the World (DCW). DCW is a useful source of country borders and topography at 1:1 million scale, and comes as a set of 4 CD-ROMs costing only £200.

DCW has been published by the US military, which has not attempted to recover any production costs. It has been agreed that there will be no further products at "give away" prices which are based on OS material.

Military Survey has been working with the UK's NATO allies to produce the Digital Geographic Information Exchange Standard (DIGEST). DIGEST has taken many man-years to produce. Although primarily produced for defence purposes, this standard is the front-runner for a European Geographic Data Exchange Standard.

3.2.4 Hydrographic Office

The Hydrographic Office is the government agency responsible for the publication of Admiralty Charts and many other products concerning the marine environment. Although the Hydrographic Office currently has a low profile within the UK GIS community, it is actively involved in the development of specialized digital chart products, in both vector and raster form. Since the Hydrographic Office has 60% of the world market for maritime navigation charts and books, its future role as a supplier of geographic data is very much an international one. The Hydrographic Office is currently collaborating with OS in developing specifications for potential paper and digital Coastal Zone map series.

3.2.5 Private Sector Data Suppliers

The private sector is becoming an increasingly important source of digital map data. Organizations such as Bartholomew and the Automobile Association have well-established track-records in cartography and have seen map data products as a natural evolution of their businesses. Other organizations such as PinPoint and The Data Consultancy provide data as part of a value-added service.

3.3 Representative Organizations

3.3.1 The Association for Geographic Information (AGI)

The AGI was formed in response to a recommendation in the 1987 Chorley Report on the handling of geographic information. AGI is funded by membership subscriptions from its 600 members which include representations from; local government, utilities, central government, GIS vendors, consultants, academic and educational institutions, research organizations and service bureaux.

The objectives of the AGI are to:

- Undertake tasks to further the use, development and awareness of GIS

- Promote the application of GIS technology

Chapter 3
GIS Organizations - Roles and Responsibilities

- Provide a forum for the exchange of information in the furtherance of GIS

- Encourage the definition of technical standards and terminology

- Co-operate/collaborate with other organizations which have an interest in GIS

- Assist and encourage research into GIS-related issues

- Educate and train people in GIS

- Publish GIS related papers and documents

- Organize events to exchange information and further awareness of GIS.

Prominent among AGI's current activities are an annual exhibition and conference, the publication of a members' newsletter and yearbook, the distribution of the Tradeable Information Initiative Metadata (see section 6.1.1) and the definition of technical standards on behalf of the British Standards Institution.

3.3.2 LGMB - Geographic Information Advisory Group (GIAG)

In 1991 the four local authority associations in Great Britain agreed to set up the Board Geographic Information Advisory Group (GIAG), with support from the Local Government Management Board (The LGMB). Its primary roles are:

- To protect the interests of local government's spatially related data

- To promote the use of GIS in local government.

The Work Programme covers basically four areas of GIS:

- Standards: GIAG has led the following projects:

    - National Street Gazetteer specifications (now a British Standard)

- National Land and Property Gazetteer specification

- Address standard specification

• Tradeable information: GIAG has recently negotiated a Service Level Agreement with Ordnance Survey. It is involved in discussions with OS about revenue earning from some local authority datasets.

• Good Practice. The following publications are available from the LGMB:

    - Cost/Benefit Methodology for GIS

    - Case Study of a GIS implementation

    - GIS Functional Specification

    - An approach to evaluating GIS for local authorities (strategy)

    - Generic and specific data models, including a review of methodologies such as SSADM

    - Application Specification

    - Advice on benchmarking GIS

    - Review of GIS implementation in local authorities - successes and failures

• Awareness. A GIS information pack is in production. It details the significance of GIS and the opportunities they offer GIAG lead officers speak at the major conference, seminars and workshops. The LGMB publishes a regular newsletter "GIS News".

Chapter 3
GIS Organizations - Roles and Responsibilities

3.3.3 National Joint Utilities Group

The National Joint Utilities Group (NJUG) was formed in 1977 with representation from gas, water, electricity and telecommunications companies. A primary concern of NJUG has been to reduce the cost and disruption caused by utilities accidentally damaging each others underground plant during excavation work. In an effort to reduce the problem, NJUG has established organizations for the exchange of plant records on microfilm.

Following the Horne Report of 1985, NJUG has been working with highways authorities under the Highways and Utilities Committee (HAUC) to prepare for the New Roads and Street Works Act which came into force in January 1993. The new legislation called for a Computerised Street and Road Works Register (CSRWR). CSRWR was perceived by many commentators as having important implications for the availability of geographic information and the use of GIS in the UK. NJUG has, however, recently considered and rejected the use in the short term of GIS as part of CSRWR on the grounds of cost.

3.3.4 AM/FM International

AM/FM International is a US based organization which aims to provide a forum for the exchange of information which will facilitate the adoption and use of AM/FM/GIS technology. Members include representatives of (primarily US-based) utilities, government agencies and other agencies. AM/FM stands for Asset Management/Facilities Management and this reflects the utilities origins of the organization which originally tended to focus on asset management.

AM/FM produces a newsletter and various technical and management publications. Among its publications is a catalogue of all known magazines, books, papers and reports published on AM/FM topics.

AM/FM International has a healthy European division with active Northern Ireland and GB participation.

3.4 User and Data Forums

3.4.1 Scottish GIS Forum

The Scottish GIS Forum contains representation of several organizations with GIS interest in Scotland, including both central and local government bodies. Initiatives undertaken to date include:

- A register of Spatially Referenced Data

- A project to standardize property development data

- Other initiatives coordinated with LGMB GIAG.

3.4.2 Rural Wales GIS Forum

The Rural Wales GIS Forum is an informal group managed within the Welsh Office. The role of the group is to increase awareness of digital datasets and to facilitate their exchange. The group has also produced records of the type, scale, format and availability of paper-based and digital geographic data.

3.4.3 Interdepartmental Group on Geographic Information (IGGI)

The UK Central Government Interdepartmental Group on Geographic Information (IGGI) was formed in January 1993 as a successor to the Tradeable Information Initiative Working Group (TIIWG). The Terms of Reference for the group are yet to be finalized but will include the following:

- Coordinate the development of and adoption of spatial referencing standards for central government

- Consider the other inter-departmental aspects of the Government's Response to the Chorley Report [6]

- Maintain a register of GIS applications within central government and consider the opportunities for sharing geographic information between departments

Chapter 3
GIS Organizations - Roles and Responsibilities

- Provide a representative central government view of charging, marketing and legal issues.

- Deal with other appropriate matters as they arise.

It is very early days for IGGI but the group is likely to become an influential forum for the central government GIS community.

3.4.4 Northern Ireland GIS Liaison Committee (NIGIS)

NIGIS is an OSNI, multi-user GIS strategy group representing the interests of all the major government and public utilities in Northern Ireland. The group influences and coordinates GIS developments within the individual partner organizations to ensure eventual integration within the overall NIGIS environment.

It is assisted in this by a dedicated NIGIS Unit, staffed by and located at, OSNI.

3.5 Professional Organizations and Special Interest Groups

3.5.1 British Cartographic Society

This is the professional body for the cartographic community (not to be confused with the other BCS below). It has a GIS specialist group. The GIS specialist group produces a quarterly newsletter. Membership is open to professionally qualified cartographers.

A particular concern of the GIS specialist group is the cartographic integrity of the output from GIS. The issue is that non-cartographers may not appreciate the limitations of source material, which can result in the output from GIS being misleading.

3.5.2 British Computer Society - GIS Special Interest Group

The British Computer Society (BCS) is a professional body for practitioners within the computer industry, and its GIS Special Interest Group provides a forum for its members to exchange knowledge and experience of GIS-related matters. The BCS GIS Special Interest Group was formed in 1989 to:

- Improve the computer implementation of GIS, including the use of geographical data, spatial data and traditional databases

- Develop the awareness and competence of computer professionals in the field of GIS

- Represent the views, ideas and interests of the group members to other bodies

- Stimulate research into the technology used within GIS

- Support the British Computer Society in the development, promotion and maintenance of Standards in conjunction with other relevant Groups.

The Group holds meetings and seminars which are sometimes jointly organized with other bodies, and publishes a regular newsletter. The group has informal links with the AGI.

3.5.3 BURISA

The British Urban and Regional Information Systems Association (BURISA) is an informal, non-profit making association for practitioners whose common interest is in the use and management of information, and in the development of information systems, for services to the public.

BURISA aims to promote better communication between people concerned with information and information systems in local and central government, the health services, utilities and the academic world through its regular newsletter and periodic conferences and workshops.

Chapter 3
GIS Organizations - Roles and Responsibilities

3.5.4 Royal Institution of Chartered Surveyors (RICS)

The RICS is a driving force behind the Domesday 2000 project. The RICS has set up a *Geographic and Land Information Skills Panel* which is one of forty core-skills panels for chartered surveyors. The panel has the following general brief in the field of Geographic and Land Information Management:

- Defining and promoting the state of the art practice

- Preparing standards and practice guidance notes

- Advancing the use of technology

- Supporting the provision of training

- Responding to requests for advice and information

The panel is currently concerned with the following priority areas:

- Guidelines for the interchange of survey information in computer readable form

- An awareness campaign concerning GIS and Land Information systems. The campaign involves the use of a RICS PC-based Land Information Systems Demonstrator at conferences, exhibitions and meetings

- Assistance in the formulation of policy for the RICS on the provision of land and property information

- Fostering the need for all surveying courses to provide IT skills training.

4 Market and Technology Trends

4.1 Market Trends

Surveys conducted during the last few years reveal dramatic change in the GIS market. Areas of particular change include:

- *A growth in the value of the GIS market*

    In spite of the poor economic climate, a recent (1992) market survey [2] supports anecdotal evidence that the GIS market is continuing to grow in value. The survey suggests that the UK market for GIS was about £57 million in 1991 and realizes the predictions made in previous years. Other surveys suggest that the value the UK GIS market is considerably higher than this. The survey also suggests that the worldwide value of GIS-related business will be $10 billion by the end of the decade.

- *Reduction in costs of systems*

    Competition between vendors (particularly hardware vendors) has forced unit prices down and allowed larger volumes to be sold. This trend has applied not only to GIS vendors, but also to companies providing related GIS services.

    Whilst GIS is currently a low-volume, high margin market with a multitude of vendors, it is likely to become high-volume, lower-margin with fewer vendors selling more standard products.

- *Strategic alliances and repositioning*

    There have been many changes within the community of GIS product and services suppliers. Many of the largest hardware vendors have reconsidered their market positions, and some have disappeared from the GIS scene. Several old alliances have been broken and new ones have been formed as software and hardware companies manoeuvre for positions of sustainable influence.

- *Growing importance of desktop systems*

  IT hardware vendors are already aware that the markets for proprietary solutions in the mainframe and mini areas are dwindling and that this decline will continue during the 1990s. The growing market will be for desktop systems and competition is already developing rapidly in these areas with the announcements of Windows NT operating systems, the IBM/Apple Pink project and the shrink-wrapped Solaris product from Sunsoft.

- *Formation of new GIS-related companies*

  There is evidence of a growth in GIS-related service companies such as consultancies and data bureaux. The emergence of a new pan-European GIS magazine *GIS Europe*, is a sign that the GIS market is beginning to have an international, dimension.

  A recent article [3] suggests that UK companies have been successful in competing in Europe, and this point is supported by survey results.

- *User community becomes more aware*

  Anecdotal evidence from GIS vendors and service companies suggests that users are becoming increasingly discerning about GIS. Whereas a few years ago many organizations were buying pilot systems, these organizations now have valuable practical experience.

- *Greater availability of base mapping data*

  Whilst the geographic data market needs much greater development, the availability of data has changed somewhat from a few years ago. Both Ordnance Surveys have made great progress with their large-scale digitising programmes, and OS has collaborated with MR Data Graphics to produce the 1:10,000 and 1:50,000 scale scanned map series.

Chapter 4
Market and Technology Trends

Private sector suppliers such as the AA and Bartholomew have also extended their range of products, and are increasingly competing with OS at small scales, and at large scales in specific geographic areas such as London.

- *Development of enabling standards*

    A significant trend is the effort now being put into the development of enabling standards at European and UK levels. In the UK, for example, it was the CSRWR legislation that led to the National Street Gazetteer specification being agreed. In Europe, CEN is increasingly involved with GIS related standards which could have very significant implications for UK public procurers.

- *Practicality of in-car navigation*

    And finally, the push for in-car navigation systems has stimulated interest and activity to produce pan-European road network data.

4.2 Technology Trends

The technology trends that are influencing the development of GIS products include the following:

- *Desktop computing power*

    The power of desktop computing has seen substantial improvements. In particular, continued advances in large scale integration are leading to much faster machines. This allied with improved display technology and better Graphical User Interfaces will promote the take up of GIS desktop solutions.

- *Re-engineering of GIS products*

    Predictions for the desktop market indicate that it will be predominantly based on MS-DOS/Windows rather than the emerging desktop versions of UNIX (such as UNIVEL). The current GIS market is however, mostly UNIX-based.

- *Software architecture*

  GIS products are becoming more dependent on standard components for the provision of graphical user interfaces and data storage. Widespread adoption of MS-DOS Windows and plans for SQL3 to provide facilities for the handling of geographic data will bring about further change in the architecture of GIS software.

- *Data query tools*

  The relatively recent appearance of GUI's for PCs and increases in PC power have enabled progress to be made in providing low-cost geographic query facilities. Several suppliers who have traditionally offered workstation-based products are now offering PC versions of their products. This is an important breakthrough in that many organizations have small needs for sophisticated spatial analysis and cartography, but have substantial needs for simple query facilities.

- *Increasing use of WANs and client/server architectures*

  The availability and take-up of Wide Area Networks is having an influence on the architecture of GIS; in particular, distributed computing environments will enable greater integration of systems across geographically dispersed organizations.

  GIS require interactive graphics which are best handled by local intelligence on the desktop. Whilst this local intelligence can be provided by "X-terminals", the proliferation of Pcs will tend to drive the market towards PC-based graphics (generally using Windows 3.x). Client/server architectures based on OSF's DCE and Unix International's Atlas will have an increasing influence on GIS products.

Chapter 4
Market and Technology Trends

- *Storage capacity*

    Technical advances in processor performance have also been matched by advances in secondary storage devices, and this is of particular importance to the viability of GIS. CD-ROMs are having an increasingly important role to play in GIS solutions, and the costs of traditional magnetic hard-disks are no longer constraining applications requiring several gigabytes of on-line storage.

- *Printing and scanning*

    The growth in the market for graphics input and output devices has stimulated competitive technical advances in this area. Large format (A1-A0) colour electrostatic printing has remained out of the reach of many departmental GIS projects, and the appearance of a large format (A0) inkjet plotter at around a quarter of the previous cost has been a welcome development. Inkjet technology has also had an impact for personal GIS, with colour A4 inkjet printers now available for around £400.

    Progress has also been made in laser printing, with falling costs and the introduction of 600 dots per inch (dpi) A3 and A4 products.

    On the scanning side, most progress has been made with A3 and A4 desktop scanners for DTP applications. GIS vendors have mixed views on the importance of such small format scanners, generally preferring CCD and contact scanners.

- *Emerging technology*

    Emerging technologies such as parallel computing, multimedia, neural networks, pen-based computers and wireless connectivity have little impact on current applications, but they are likely to be important in improving usability of systems and in delivering advances in the fields of environmental modelling and geographic feature recognition.

- *Querying facilities*

    The provision of a natural language (plain English) querying facility [8] may be expected to encourage the take up of GIS technology by smaller departments and agencies lacking an adequate support base of SQL expertise.

5 Influencing Factors

5.1 Introduction

In 1983, the Shackleton Select Committee on Remote Sensing and Digital Mapping reported that the range of anticipated potential application areas within Central Government included:

- Mapping and analysis of census data
- Police and fire services
- Land registration
- Land use changes and habitat monitoring
- Land suitability classification
- Rivers, water supply and sewerage networks
- Forest management, yield forecasting, tree censuses
- Environmental mapping
- Geological mapping
- Geography-based research
- Public information systems via TV
- Highways engineering and management
- Geography-based planning
- Weather radar
- Defence requirements
- Asset management.

Nevertheless, many of the expectations of the Shackleton Committee and the subsequent Chorley Report[5] have not been met. Whilst there are a number of examples of GIS in each of the above applications areas, its potential remains largely untapped. Government use of GIS is still in its infancy.

In the following sections we intend to explain why this is the case and to establish a basis for our recommendations.

5.2 Triggers

Use of GIS is being triggered by changes in organizations' business activities and by certain major projects. New GIS initiatives are very often the result of changes in business activities. In particular, departments are often asked to undertake new duties as a result of emerging legislation or as part of service improvement. As consideration is given to the working practices involved, GIS is often identified as a prospective business tool. Examples of legislation that can be supported by GIS include:

- The introduction of a property-based Council Tax
- The introduction of charges for fishing rights
- The Roads and Street Works Act
- Requirement for Coastal Zone Management.

Changes in business activities do not only occur with the introduction of new working practices. Organizational changes can stimulate a review of how a business operates. For example the formation of an Environmental Agency is expected to see a consolidation of existing pollution control activities into so-called Integrated Pollution Control.

Natural and man-made incidents are also stimuli to changes in business activities. For example the successful application of digital map data in the Gulf War has led to greater recognition of its value for military operations, and GIS played an important part in the assessment of the Exxon Valdez oil spill.

Chapter 5
Influencing Factors

Project-based GIS
GIS also plays an important part in one-off projects such as research projects or major engineering schemes. In some cases GIS is simply seen as an efficient way of working, in other cases it is vital technology. For example, in 1987 Sir William Halcrow & Partners were appointed to undertake work on a National Sea Defence Management Study, and this project used GIS technology to record, manage, analyze and present the findings.

British Rail has made heavy use of GIS in planning various options for the High Speed Rail Link from London to the Channel Tunnel.

5.3 Drivers

There are four basic forces which drive organizations to use GIS. These are:

Improving levels of service
There is widespread agreement that GIS can be used for improving organizations' levels of service to their customers. GIS is able to do this through:

- Improving the quality of data presentation

- Providing faster access to information

- Providing easier access to information

- Making information more usable

- Allowing geographic information to be integrated with "conventional" data

- Improving data integrity through new methods of information handling

- Allowing new forms of information to be produced

- Raising the image of the organization

- Sharing of information gathering with, and of information held by, other organizations.

Reducing operating costs - efficiency
Just as database technology has transformed the efficiency of handling textual and numeric data, GIS has had a similar impact on geographic data. Quite often GIS can be more labour intensive than manual systems when it comes to data collection - it is with data maintenance, data use and data retrieval that significant staff efficiencies are obtained. An analogous situation applies, for example, to conventional databases and wordprocessors. Whereas business managers rarely question the use of a computer database as opposed to a card index, or a wordprocessor as opposed to a typewriter, there is often deep suspicion concerning the use of GIS instead of paper-based systems. However, it should be stressed that GIS is not a panacea, and just as there remains a role for the card index and the typewriter, GIS is unlikely to eliminate the paper map from modern businesses.

The lack of documentary evidence of improved staff efficiencies is due in part to the fact that GIS changes the nature of working practices to the extent that it is very difficult to compare like with like.

From a mapping viewpoint, the base of cartographers with traditional skills is diminishing. Recruitment could be a problem in future. GIS offers some relief from this problem.

Better Targeting of Resources
GIS is often applied to improve the targeting of resources by delivering tangible improvements in the quality and variety of information available to business managers. If one considers that an estimated £30 billion of UK funds are targeted with information produced by OPCS alone, - what is the value of being able to produce new forms of data? - or of making the data available more quickly? Arguably a modest 1% improvement in targeting these resources should be worth £300 million.

Using GIS as a critical technology
In some cases using GIS technology is the only feasible way to perform certain tasks, and, provided the task is important enough, this can be the overwhelming decider that GIS is to be used. For example, monitoring national land-use change is practically impossible to do using field work techniques, but is viable by adopting satellite imagery techniques.

5.4 Enablers

There are a number of developments and trends that are making GIS exploitation more likely.

Technical breakthroughs
Although some forms of geographic and cartographic systems have existed from the early 1960's, it was a series of technological breakthroughs beginning in the early 1980's which were critical in enabling modern GIS products to be produced:

- Raster graphics workstations - graphic processors, larger memories, faster processors

- Graphical user interfaces - WIMP interfaces, graphics libraries

- High capacity on-line storage - CD-ROMs and magnetic disks

- Scanning systems - CCD and contact scanners

- Raster output devices - laser, electrostatic, thermal wax and inkjet

- Networks - LANs and WANs which have enabled multi-user systems to be produced

- Software advances - which have to some extent made system development easier.

Inevitably further technical advances will occur and these will have a positive effect on GIS; the general consensus, however, is that the pace of technological change far outstrips demand.

Financial breakthroughs
Due to an increasingly competitive and commodity-style market, the cost of hardware, software and services has fallen in real terms. It is thus becoming easier to make good financial cases for GIS projects.

Improved availability of geographic data
In the early days of GIS there was an absence of base map data. For many applications, the availability of base map data is less of a problem today. Some pundits have argued that the UK now has the critical mass of base map data it needs for a viable GIS market. However, from an individual organization's point of view, the availability of data is closely associated with the data pricing issue; indications are that the cost of data is hampering market expansion.

Aside from base map data, some progress has been made in increasing the volume of available foreground data. This has come about by the release of private sector products and a small volume of digital map features from public sector user departments.

Awareness
Awareness of GIS and an appreciation of its role within information systems strategy is a vital enabler. Currently business managers have little difficulty in recognising business applications which require conventional computer technology - in fact many organizations often ignore the option of manual systems for some applications. Yet when faced with new working practices or projects involving geographic information, the option of using GIS technology is often either not considered or is regarded as not yet practical, and passes by as a lost opportunity.

The Association for Geographic Information intends to change this situation - not by advocating GIS technology per se, but by educating senior managers to recognize when GIS should be considered.

5.5 Barriers

Although the reasons for investing in GIS are quite attractive, many organizations find the reasons for not implementing it even more compelling.

5.5.1 Data

Awareness of data availability
There is a widespread lack of awareness of data availability and plans to make geographic data available. Obtaining such information as a prerequisite for establishing the feasibility of GIS projects is a problem.

The Tradeable Information Initiative Metadata has been a step forward in that it catalogues many existing geographic and spatially-referenced datasets; however, interviewees identified several problems with it:

- Usability of the information in the form issued
- It will quickly become out of date
- It does not include planned data availability.

Data availability and cost
Whilst substantial progress has been made in making new base maps available, this still represents a major barrier for some organizations and applications where digital map coverage is non-existent or incomplete. The same concern applies for foreground datasets. The complete or partial absence of digital data in one organization hampers the development of applications in other organizations. As noted earlier, the cost of geographic data also represents a barrier to some organizations.

Data quality
Many parts of government are concerned about the quality of data, including its:

- Accuracy
- Completeness
- Currency
- Scale.

The integration of GPS-based systems with existing base map data can pose major problems. Data quality is an issue for departments from both a data purchaser's and a data supplier's perspective.

Data standards
Although there has been much recent activity on data standards, until this work begins to filter through into data products and systems, it will remain a problem area. The areas of data standards which apply to GIS are:

- **Spatial referencing.** This will identify a definitive set of basic spatial units (BSUs), their owners and the community of users for this data

- **Data quality.** This includes the production of appropriate "generic" quality assurance and control definitions, guidelines and indices

- **Data transfer.** In the UK this has primarily focused on the development of the NTF standard which has recently been published as BS7567; DIGEST (see 3.2.3) is the prime candidate for a European Geographic Data Exchange Standard

- **Standard data models.** This includes the street gazetteer, land and property gazetteer and models for specific uses of transfer mechanisms such as the NTF.

Several of the people who were interviewed during the study leading to this report felt that international standards were of greater importance than British standards. Public procurers must of course adhere to relevant international standards, where these exist.

Chapter 5
Influencing Factors

Data conversion
With the exception of a few organizations that are developing second-generation systems, GIS projects generally involve a substantial volume of data conversion. Data conversion programmes are regarded as a major hurdle, primarily due to the fact that they can account for up to 80% of the project costs. They also raise issues relating to data quality, particularly when they draw together information from disparate sources.

Release of Data
The release of data is a major issue amongst organizations which are already engaged in GIS activities. Quite simply organizations often have little or no experience of it, and regard it as a complex subject. Furthermore, since they are generally under no pressure to release it anyway, they choose not to pursue it. Ironically these organizations often complain about the absence of key data sets from other organizations.

The issues which concern data release include:

- Pricing — What should be charged? What about the EC directive on Freedom of Environmental Information?

- Security and Privacy — Does its release have national security and implications? How can any required confidentiality of information be safeguarded? How can unauthorized modification of the data be prevented? How can access to prevent access be prevented? What are the rules on privacy? How does the Data Protection Act apply?

- Sensitivity — Is the information sensitive? Could it affect property prices?

45

- Copyright — What about derived data? How can copyright be enforced?

- Liability — Are there liability issues? What if the data is wrong?

- Quality — How can quality be measured and indicated?

- Ownership — Who owns data collected by a third party?

- Method — What forms of release are there? What are the pros and cons?

5.5.2 GIS Implementation

Market Immaturity
The GIS market is still perceived to be immature and there are a number of problem areas associated with this fact:

- Suppliers have over-hyped their capabilities and raised expectations

- System selection still involves risks concerning the suitability of the systems proffered and uncertainty concerning the medium- to long-term future of the company involved or the company's commitment to GIS

- The systems have some way to go in terms of usability.

Technical Complexity
GIS is seen to be technically complex technology, requiring specialist skills to procure, implement, manage and use.

Culture Change
Users often feel threatened by its introduction - particularly if they are unskilled. Training is essential to raise user confidence.

A culture change is usually required to accept that most geographic information is a corporate resource to be shared. Because GIS often breaks down business area barriers, high-level vision and commitment are generally needed to support the changes involved.

Running pilots in parallel
Running a pilot project in parallel with normal operational activities, though good management practice, is often a disincentive to adopting GIS since it either requires additional resources or places existing staff under greater pressure.

5.5.3 Implementation of Policies

Market Testing, Executive Agencies and organizational fragmentation

Market Testing has produced uncertainty in some government departments and agencies, which has in turn brought about a reluctance to invest in new information systems, including GIS.

Ironically, some of the new information systems which are currently being commissioned are focused on contract management which, for certain types of contract, should be an opportunity for some forms of GIS. Many such opportunities are lost due to lack of awareness.

One particular impact of Market Testing is that it has raised concerns about how to deal with long-term initiatives and projects, such as GIS, when service agreements and contracts tend to be relatively short. The contracts can, of course, be made as long as needed, but especially in relation to long term projects, particular care needs to be taken over departments strategic control of IS.

Departments are also unclear about how treating geographic information as a corporate resource will work as their organization is fragmented. This issue also raises questions concerning how OS will treat such organizations and who owns the data.

An interesting case arose when the Thames Water Authority was split into Thames Water PLC and NRA Thames Region. In the case in question, the OS position has been that NRA should pay for the purchase of digital maps, in spite of the fact that no new use of the maps was being made, as the maps were provided under a Facilities Management (FM) agreement with Thames Water (NRA does not actually have copies), and the Act of Parliament which created the NRA in September 1989 gave explicit rights to share such archives. This sort of problem may now proliferate and has no clear resolution.

Cost Recovery
Cost Recovery is another policy which has been cited as a barrier to the use of GIS. Whilst the policy is clearly successful in creating a business-like climate within government, it has put in place tariffs on information which has hitherto been shared freely or at marginal cost. With financial pressures ever-present, there is a fear among some interviewees that Cost Recovery could produce an "information recession" as departments try to balance their books.

A further aspect of Cost Recovery is that many government organizations have to obtain special agreement from HM Treasury to retain funds generated from data release rather than have them go into the Consolidated Central Fund. Some interviewees asserted that the inability to retain revenues in the business reduces the incentive to release data, either internally to other parts of government or to the private sector.

OS Policies
OS policies were cited by interviewees more often than any other barrier, and it has been the subject of great debate within the GIS community. The Association for Geographic Information provided a forum for round-table discussions in October 1992 as a result of widespread concern among its members.

Chapter 5
Influencing Factors

Key points arising from the round-table discussions were concern about:

- The level of charges set by OS

- The stability/predictability of OS charges

- The charging mechanisms

- OS use of copyright to block competitive products

- The nature of some OS products.

OS are now, in general, moving towards a policy of leasing rather than selling data, in order to achieve a regular revenue stream over a period of years. The OS policy objective is primarily to maximize use, subject to meeting financial targets.

OS have two key cash-flow problems:

- The need to recover money quickly, ie within the financial year for ever increasing cost-recovery targets

- They have to justify investment in long term projects, where benefits will not be realized for many years.

The OS are concerned that cost comparisons with their pricing are not made on an equitable basis - the cost of digitising map-based records elsewhere in government may include only the *marginal* cost of staff time, whereas OS are tasked with recovering the whole cost of original data capture, cartography, fixed overheads and maintenance. Although the historic costs of the survey of large scale mapping are regarded as 'sunk' costs, OS is moving towards full recovery of in-year costs. In 1993/4, OS is targeting 70% cost recovery.

The fact that departments may apparently find it cheaper to re-digitize key features even though OS may already have them in digital form is a serious concern.

Interviewees also stated that:

- OS has spent millions of pounds in a programme to capture large scale digital data on the basis of unwritten promises of purchases which in some cases have not yet materialized

- OS has needed to change the specification of its products to meet user needs. The current absence of a topological structure poses questions about whether most OS data fully meets users' needs

- OS appears to have been hesitant in producing relatively high-demand products such as 1:50,000 and 1:10,000 scale scanned maps which are widely applicable and less expensive to produce than vector data. OS has embarked on producing both sets of scanned products as a collaborate venture with a private-sector partner.

OSNI Policies
The OSNI digital product has always included features such as multi-attributes, links and nodes, edge-matching and associated textual data, as a necessary pre-requisite for its role in underpinning NIGIS.

An extensive review of OSNI charging policy has just been completed and awaits acceptance by Treasury. While copyright will continue to feature to some extent, in general, a low cost pricing policy for digital data has been recommended as essential to the long-term viability of NIGIS.

5.5.4 Information Systems Strategy

Mutual awareness
Some interviewees felt that there is a general lack of awareness among departments of their respective IS-related activities. It was felt that many opportunities for collaboration on large projects such as GIS and multi-departmental business systems were lost due to a lack of mutual awareness. For GIS projects, awareness of the intentions of other government bodies mainly concerns mutual interests in geographic information.

Geographic Information in the IS Strategy
Another barrier is that departments generally have difficulty in, or are sceptical about, placing GIS within their IS strategies, and often set it to one side as an isolated system or leave it out altogether. In the absence of a coordinated framework in which GIS can be exploited, individual projects are faced with justifying shared infrastructure components such as base maps, data capture, printing facilities, and core GIS software. This reduces the strength of business cases for individual applications when often a collective case could be made.

A large proportion of GIS opportunities will not be realized unless GIS becomes an integral component of long-term IS plans.

5.5.5 Awareness

Senior Management Awareness
Some interviewees suggested that current levels of awareness are a major barrier to the growth in use of GIS. The levels of interest at conferences and exhibitions suggests that awareness of GIS is growing, but the feeling among interviewees is that awareness is a problem at senior management and at policy-making levels in government.

The AGI has been active in addressing senior-level awareness through events and, for example, has hosted Parliamentary IT Committee (PITCOM) meetings. The AGI is concerned not only that the right people have heard of GIS, but also that prospective users of GIS are given a consistent and objective message. The current feeling is that vendors and enthusiasts have over-hyped GIS capabilities, often leading prospective users to expect too much, and sometimes leaving them confused or sceptical.

5.5.6 Costs and Benefits

Project Costs
GIS projects are still regarded as expensive - particularly for first-generation projects needing large data conversion programmes. Even when a long-term financial case can be made for GIS, many organizations are unable to allocate the resources required.

Lack of proven benefits
Currently, senior executives are reluctant to implement GIS projects because they find it difficult to justify the investment, especially if the initial case indicates several years before the project produces a financial return. GIS suffers from the problem that its biggest benefits are sometimes intangible, and when business decisions are made on financial benefits alone, these may not be sufficient.

A major source of difficulty for potential user departments concerns qualifying and quantifying the benefits. One approach, based on actual project experience, was developed by Harold Thurman in the US, following his review of costs and benefits of the use of Digital Map Databases in North America, known as the Joint Nordic Report. In a paper to the 1990 AM/FM International Conference he identified fifteen potential sources of savings and benefits.

Such findings could be used as a basis to determine the potential benefits of GIS in other organizations. Basing business cases on actual experience can allow intangible benefits, which would otherwise have remained outside the business case, to be quantified.

The lack of recognized documentary evidence of this sort, based on UK experience and prices, presents would-be project managers with difficulties in establishing the credibility of their case.

6 Opportunities for Working Together

Preceding sections have briefly identified a variety of factors that have limited the use of GIS for business purposes. As part of the study, the consultants engaged by CCTA also examined possible methods of overcoming these barriers. Whether or not some of these ideas are viable remains to be determined. However, they are all included in this report in the interests of presenting a full picture of the consultants' findings. In several cases interviewees have suggested initiatives which have been found to be already in place.

Chapter 7 recommends some of the ideas considered to be most needed and appropriate for CCTA to undertake; other ideas are left as suggestions for the AGI, IGGI and individual parts of government.

6.1 Data and Systems

6.1.1 Metadata

The former Tradeable Information Initiative Working Group (TIIWG) produced a register of information sources (metadata) within government, and this is currently distributed through the AGI in machine-readable form.

Several interviewees felt that the metadata has raised mutual awareness considerably, but there was concern that the metadata will need to be maintained and enhanced over time. It has been suggested that a simple geographic package, such as the British Oceanographic Data Centre's Digital Marine Atlas, would be an appropriate way of illustrating the extent of geographic data holdings.

The Inter-departmental Group on Geographic Information will be considering the future of metadata.

6.1.2 Data standards

Many interviewees felt that there was a need for work to be carried out in the field of data standards. Data standards are mainly required to reduce technical risks, and to reduce the need for bespoke development.

The AGI Standards Committee is already undertaking a programme of work to develop data-related standards. The *AGI Scope and Strategy for GIS* identified the following areas for data standards development:

- Spatial units
- Data quality
- Data transfer mechanisms
- Data models such as street gazetteer, land and property gazetteer

Current standards activities, led or co-ordinated by the AGI, are addressing:

- Spatial Unit Definitions
- Spatial Unit Boundary Data
- Spatial Unit Attribute Data
- Spatial Referencing
- Quality Model
- Quality Assessment of Data
- Quality Assurance
- Spatial Data Types and Standard Data Models
- Street Gazetteer specification
- Land and Property Gazetteer specification
- Address specification
- Conceptual Data Models.

Chapter 6
Opportunities for Working Together

On this evidence the AGI standards programme is adequately covering data standards for geographic information and GIS. However, the AGI will need some external finance to carry out this large programme of work.

6.1.3 Systems Standards

There is a requirement for the development and implementation of systems standards to enable GIS tools to coexist with other systems. The AGI has undertaken a range of tasks under the auspices of its Standards Committee. These include the publication of the report *The Scope and Strategy for GIS standards [4]*. This document argues that the absence of standards adds significantly to the technical risks and costs of GIS projects.

The Scope and Strategy for GIS Standards justifies the adoption of standards on the basis of:

- The ability to use standard hardware platforms

- The ability to use standard software tools

- The reduction in required systems development effort

- The need for phased implementation and integration of systems

- The need for portability, scalability and inter-operability and choice of system supplier.

The report goes on to suggest that GIS-related or GIS-adapted standards should be developed in the following areas:

- Applications development methods, tools and languages

- Human-Computer Interaction

- Data interchange and access, based on SQL, IRDS and similar standards

- Systems management and administration.
- Benchmarking standards.

It will also be necessary for conformance tests to be defined and applied to ensure that claims to meet these standards can be authenticated.

Work already undertaken in this area by the AGI includes:

- Development of a "Terms of Reference" for GIS implementation in a de facto/de jure standards based environment

- Contribution to the development of the SQL3 standard

- Proposals for a standard GIS User Interface and input to the ISO GUI standards committees

- Input to the BSI IST/20 IRDS Standards Committee to discuss the way in which geographic information will be described, handled and stored

- Development of conceptual data models for data transfer standards and contributions to the review of ISO 8211 which defines a generic data transfer format

- Review of a directory of existing and proposed standards for the UK only.

The AGI Standards Committee plans to develop a "road map" of emerging standards which are relevant to GIS and to produce guidelines on de facto standards.

6.2 Implementation of Policies

Again it appears that work undertaken or planned by the AGI will adequately cover GIS-related systems standards. However, funding levels are such that progress in developing those standards is rather slow.

6.2.1 Review of the impact of Market Testing and Cost Recovery

Government Market Testing and Cost Recovery policies are intended to reduce the costs of operational services in departments and agencies, but their impact is not confined to the balance sheet. The study consultants found widespread concern however about the effects of these policies on government information. In particular it was felt that information would become increasingly expensive to obtain and use as organizations become fragmented, and a commercial ethos reduces the willingness of the departments and agencies to provide information to each other.

Market Testing seeks only to compare the current in-house IS services for effectiveness and value-for-money against alternative sources of service provision. However, despite advice to the contrary, for example CCTA's Market Testing publications, Market Testing has apparently produced short-term reluctance to invest in new systems, and a less certain medium- to long-term future for particular service providers. For projects that require medium- and long-term planning, as is often the case with GIS, this has the potential for becoming a substantial obstacle unless the matter is dealt with. In particular the onus falls more than ever on the business areas to control the development and exploitation of IS in general and GIS in particular for the benefit of their businesses, irrespective of who provides the IS service.

Similarly, the Cost Recovery policy, coupled with the creation of smaller business units, has sought to focus departments' attention on value for money, and information provision policies are often based on commercial-style, organization-specific benefits. Many interviewees felt that information provision, in respect of either GIS or conventional information, should, in some cases, be examined from a wider government perspective, with more emphasis placed on coordination for the greater good. Government organizations are, for example, tending to be very selective about the data they gather for use - ignoring perhaps relevant, but non-essential information in order to ameliorate their financial position. This could increase the cost overall to the tax-payer.

Equally there is evidence that some government organizations are reluctant to purchase information from other government organizations, since they can purchase equivalent information more cheaply from the private sector. Yet, from a certain perspective, the government has already paid for the information to be gathered and is arguably, therefore, faced with paying twice.

The current application of these policies is raising concerns among some interviewees and GIS commentators that government could find itself in an "information recession" as government organizations fragment and tariffs on information begin to have an effect. It has, therefore, been suggested, by various interviewees to the study consultants, that a review of the impact of these policies on information exchange should be undertaken, with a view to addressing the issues and risks cited in this report.

6.2.2 Guidance on data release and purchase

The general feeling among interviewees was one of considerable uncertainty concerning the release of their geographic information and many expressed a need for guidance in this area before real progress can be made. The uncertainty surrounds the application of policies and legislation which have combined to become a strong deterrent to data release and purchase.

Among the issues which interviewees requested should be addressed in such guidance are:

Cost Recovery
- What is the thinking behind Cost Recovery?
- How does it affect geographic data pricing?
- Is it possible to retain income from data sales?
- What if there is a competing product?

Tradeable Information Policy
- What is the Tradeable Information Policy?
- How does it work?

Chapter 6
Opportunities for Working Together

- How does the Tradeable Information Policy relate to Cost Recovery and Market Testing?

- Case studies of trading government-held data through private sector companies.

Freedom of access to information on the environment.
- How do we interpret EC Directive 90/313?

- Does it conflict with other policies?

- What does it mean in terms of data pricing?

Market Testing
- How do we plan for GIS with Market Testing?

- How do we determine data ownership?

- How do we specify a service contract for GIS facilities?

- If we contract-out data collection, who will own the data produced?

Copyright and data ownership
- How do we determine data ownership?

- What are the principles behind copyright legislation?

- Does Central Government own data provided by local government?

- What are the implications for derived data?

- How can copyright be enforced?

- What if the private sector produces similar products?

- Where can we get advice?

Liability, Privacy & Security
- What are the rules on liability, privacy and security issues?

59

- Why is it considered acceptable to identify the owner of a piece of land, but not to identify the land owned by a particular individual?

- What will be the effect on property prices if the contaminated land register is public knowledge?

- What if the released information is wrong or misleading?

Quality
- How do errors propagate?

- How do we specify accuracy, completeness, currency?

- How can we determine whether a specified quality is good enough?

Data Standards
- What standards currently apply in the UK, EC, and beyond?

- What will the future standards be, and how can we influence them?

- What should we do if it costs more to adopt data standards?

Practical guidelines for dealing with these issues are essential before many parts of government will have the confidence to release their geographic information.

6.2.3 Review of government data exchange

One specific area of concern was the interface between local and central government. Local government is often tasked with providing data to central government, but since central government departments are largely autonomous units, information is requested in a wide variety of forms. This problem also exists between central government departments and Non Departmental Public Bodies (NDPBs). The inconsistency of information requests inevitably complicates the compilation and delivery of the information. It was suggested to the study consultants by a number of the interviewees that there should be a review of information flows between local government, central government and NDPBs in the interest of adopting a consistent approach.

6.3 Strategy

Whilst it is common practice for individual government bodies to define IS and GIS Strategy plans to suit their own requirements, there is currently very little inter-departmental coordination. Generally, multi-partner projects such as Domesday 2000 are only viable if there is substantial commitment from all organizations involved.

Several interviewees felt that this interdependency made it difficult to plan large or long-term GIS projects, and that some level of coordination and consistency was desirable. There were two specific suggestions made to ease this problem:

- Guidance on GIS Strategy, and
- Exchange of GIS/IS Strategic plans.

6.3.1 Guidance on GIS strategy

Several interviewees made reference to the lack of experience in planning IS Strategies, and associated policies, which involve GIS and suggested that some guidance was needed. There are several issues which this involves:

- Communications infrastructure
- Systems architecture
- Data sources
- Integration with existing IS
- Implementation policies
- Using GIS to achieve corporate objectives.

A common problem is that GIS is often treated as a business application in its own right, and is set to one side as a self-contained system, whereas, for best effect, it needs to be established as an integral part of an organization's IS infrastructure and to be applied according to business needs. Until GIS is correctly identified in IS Strategy plans, it will remain a poor relation to other IS facilities even within organizations in which geographic information is dominant.

6.3.2 Exchange of IS/GIS Strategy plans

Whilst guidance on GIS within an IS Strategy may well improve the quality of IS/GIS strategic planning, it will not necessary lead to high-level coordination. Since individual departments define their IS strategies according to departmental needs, imposing a coordinated approach would be counter-productive. It would however make sense to increase mutual awareness, and one suggestion is that departments should exchange IS/GIS Strategy plans. This would appear to be a positive step for a variety of reasons:

- It may help raise standards of GIS-related IS planning

- It would enable individual departments to learn from approaches adopted in other departments

- It would raise awareness of opportunities for collaboration

- It would help define "best practice" and encourage its use

- It would help improve the viability of long-term GIS projects.

6.4 Awareness

6.4.1 Management Publications

Awareness has been a problem for GIS ever since its conception, and whilst much progress has been made in increasing the number of people who are aware of its existence, the main area which needs to be addressed concerns the level of their understanding. A commonly held view was that lack of properly informed opinions on GIS is most acute at senior levels.

Another widely held opinion is that GIS has been over-sold and this has led to numerous misconceptions about its capabilities. Not surprisingly, business managers are conscious that they have been exposed to exaggerated claims and are more cautious when it comes to making a decision.

There is a clear need for an objective explanation of what GIS is and why it is relevant to government. This should be approached from the point of view of business needs, rather than over-sell with a technology-led sales pitch. During the course of this study, CCTA has funded the development of a publication "Introduction to GIS for Business Benefit".

6.4.2 Case Studies

Another common request was for documented case studies. These are required for:

- Evidence of achieved levels of benefits of GIS

- Information on:

    - alternative technical solutions

    - product capabilities

    - potential problem areas

Whilst publications such as Mapping Awareness and AGI Year Book carry descriptions of projects, these are often biased to suit public relations needs and to maintain the interest of magazine readers. The need for case studies extends beyond a superficial level and into the detailed history of the methods and technology used and how decisions were taken.

LGMB are currently preparing a generic local government case study to focus on this sort of information.

6.4.3 Other Ideas

A few other ideas were suggested for raising awareness, and these were aimed to overcome the fact that GIS is difficult to appreciate unless you have seen it.

- A video could be produced to show how geographic information is currently used, and to demonstrate what GIS could do to improve working practices. There are several videos currently available from vendors, but these tend to be promotional rather than educational.

- Demonstrator software could be distributed to allow prospective users to determine for themselves whether GIS is relevant to their business. There are already a small number of public-domain demonstration packages available which could be used for this, and several suppliers produce promotional "slide-shows".

- A variation on the above idea would be to use multi-media tools to integrate sound, video and computer graphics.

A seminar for senior managers dealing with the relevance of geographic information and GIS to the businesses of government organizations was hosted by the Department of the Environment, Ordnance Survey and CCTA in April 1993.

6.5 Organizational

6.5.1 Inter-departmental data forum

Opportunities for inter-departmental cooperation on data supply exist for many forms of data, and this is particularly true of geographic datasets. One approach which was prominent during the study was the formation of an interdepartmental forum for geographic information. The Scottish GIS Forum and the Rural Wales GIS Forum illustrate the demand for this.

Such a forum could be an informal meeting point for representatives from organizations engaged in GIS activities. By increasing mutual awareness,

Chapter 6
Opportunities for Working Together

activities. By increasing mutual awareness, opportunities for collaboration and coordination could be identified, and responsibilities undertaken on a voluntary basis. For example, where two or more departments have a requirement for the capture or conversion of a geographic dataset, it may be possible to share the costs among participating organizations to lessen the financial burden on each.

Such a forum could also be used to obtain agreement on issues concerning the specification of:

- Quality of source records including currency, accuracy, scale and completeness

- Content, structure and format of the data produced

- Spatial referencing method

- Timescales for availability

- Data capture techniques.

The Interdepartmental Group for Geographic Information (IGGI) was formed during the course of the study, and is certainly expected to serve much of this need for many parts of government. Notably, the membership of IGGI was a topic for discussion at its inaugural meeting, and it has been confirmed that Non Departmental Public Bodies will be allowed membership provided individual cases are justified.

6.5.2 Central Government Representation

A more formalized body could be formed to undertake tasks which are of wide application or are of particular importance to government. Whereas local authorities and public utilities have dedicated GIS groups within LGMB and NJUG respectively, there is no equivalent for many parts of government. The fact that NJUG and LGMB have resources specifically allocated for GIS-related issues has meant that they have been able to undertake activities which would have been difficult to achieve through voluntary effort alone.

The general feeling in government is that the AGI should undertake many of the new initiatives, yet there is a general apathy when it comes to providing the AGI with the sponsorship it needs.

6.5.3 Geographic Information Archive

Since the release of geographic information is potentially complex, it has been suggested that an organization could be nominated to undertake data release for the whole government community - a Centre for Geographic Information. This idea may have particular relevance in the current climate of Market Testing.

There are many attractions to this idea in terms of:

- Complex copyright, liability, pricing and security issues could be dealt with more efficiently

- It would allow "one-stop shopping" for geographic data.

A major difficulty with this idea is that it relies on a multi-departmental agreement and business case in order to have a future.

Another problem would be the risk of breaking existing arrangements on the part of individual government organizations for the supply of geographic information through networks of agents and outlets. The market for marine navigational data, for example, is different from the mainstream GIS market and might have to be treated as one of a number of special cases.

A simplified implementation of this idea would be to set up a single enquiry point with expertise to direct people to the right source of data and to monitor the service provided by the sources.

In terms of the public sector performing such a role, the Ordnance Survey and British Library stand out as organizations with an ideal background and experience in this area. However, the private sector could also play a role in this. The DTI's guidelines on Tradeable Information set out how government can work with

Chapter 6
Opportunities for Working Together

private sector organizations to release data, and there are a number of examples of government bodies using private sector companies in this way.

6.6 Implementation Management

6.6.1 Buyer's Guide

There was general agreement amongst interviewees that GIS suppliers have over-sold their capabilities, and this also comes through in system proposals. A number of interviewees felt that it should be much easier to buy from suppliers with proven capabilities. Whilst it would be unfair to offer subjective guidance on the suitability of individual suppliers, it was agreed that an objective description of GIS suppliers and their products would be of great value. During the course of the study CCTA have produced such a guide.

6.6.2 Business/Systems Analysis Methods

Several interviewees indicated that there was a need for business and systems analysis methods to be enhanced for geographic information. Areas of difficulty include:

- Data modelling of geographic features

- Data flow

- System sizing

- Performance specification

During the course of this study CCTA has funded a study to investigate the application of SSADM, their system development method, to GIS.

7 Conclusions and Recommendations

7.1 Conclusions

Geographic information is a fundamental resource within government businesses, and GIS offer the opportunity for the full potential of such information to be realized. Working in favour of wider adoption of GIS are factors such as:

- Anecdotal evidence of quality of service benefits and efficiency savings

- GIS vendor community claims

- Business imperatives which demand the use of GIS.

However, many parts of government have indicated that they are finding it difficult to make a sufficiently convincing business case because:

- GIS is still regarded as technically complex and risky - the market is established but is still immature. Business managers need to have high levels of confidence in projects before they will commit to them

- IS Strategies often ignore GIS altogether, or set it to one side as a specialist system. GIS needs to lose its specialist status and become established as just another business tool among a range of others. Its application should be business led - not technology led

- There remains a lack of awareness and understanding of GIS at senior levels. Opportunities for its use often pass by unnoticed

- Generally GIS projects are expensive and time-consuming. Even if a financial case can be made, organizations often have more pressing priorities for their financial resources and staff

- There is a lack of documentary evidence of the benefits of GIS, and this presents would-be GIS projects with a credibility problem

- Obtaining essential data is often difficult and expensive, particularly for first-generation uses which require substantial investments in data conversion and purchase

- Cost Recovery has introduced tariffs for information flowing within government, and may be discouraging both its release and use

- Market Testing has created a climate of uncertainty regarding information systems, and in some cases this is discouraging the long-term planning necessary for many GIS projects

- The unstable and, some would say, high prices of data products from OS are dampening the take-up of GIS.

In short - for many parts of government, the arguments in favour of GIS will not be compelling enough until a number of obstacles are removed and the pressure to adopt GIS comes from genuine business priorities.

7.2 Recommendations for further work by CCTA

The widely-held view of interviewees and that of the consultants engaged in the study, is that GIS technology **does** address genuine business needs and that CCTA should continue to engage in activities which advance the realization of its potential.

Some of the interviewees' concerns are already being addressed through the work of AGI, IGGI and CCTA. Recommendations for further work by CCTA are given below.

Chapter 7
Conclusions and Recommendations

Recommendation 1 - Business Benefits of GIS

It is recommended that CCTA undertakes a study to document the business benefits of generic GIS applications in order to provide a firmer basis for GIS business cases. The study should be scientifically based, objective in its approach, and should as far as possible obtain its findings by examining genuine cases.

Recommendation 2 - Impact of Market Testing, organizational fragmentation and Cost Recovery on government information exchange.

It is recommended that CCTA undertakes a review of the impact of Market Testing, organizational fragmentation and Cost Recovery on the inter-departmental exchange of information. The study should examine the impact of charging for information and make comparisons with practice overseas.

Recommendation 3 - Produce guidelines for government data release and data purchase

It is recommended that CCTA should produce guidelines to help government with the release and purchase of geographic information. The guidelines should make reference to current and proposed UK and EC legislation and should define a practical framework around which departments can develop data release services with confidence.

It is assumed that IGGI will resolve the problem of departments and agencies not being aware of each other's plans for the holding of geographic information.

Recommendation 4 - Guidance on GIS within IS Strategy

It is recommended that CCTA should either produce new guidelines on how to accommodate GIS in IS strategy plans, or should review current IS Strategy publications with the same objective. This work should take cognisance of the emerging guidance from CCTA on information and data management.

Recommendation 5 - Continued GIS involvement

It is recommended that CCTA maintains an active role in GIS-related activities and should use its position of influence within government to raise GIS awareness and understanding. In particular CCTA should:

- Become a sponsor member of the AGI

- Consider how it might facilitate the exchange of IS Strategy plans to encourage awareness, collaboration and coordination

- Contribute to the urgent development in accordance with government organizations needs of standards and methods; and their promulgation within government

- Monitor the need for a body to represent government GIS interests

- Take the lead in creating short-term groupings of interested parties to address specific GIS-related issues affecting government organizations, as they arise.

Glossary

AGI	The Association for Geographic Information. The UK umbrella organization for geographic information and its associated technology.
AM/FM	Asset management / Facilities Management
Application development	The development of bespoke application software.
Atlas	Proprietary product of Unix International.
Background maps and Foreground Features	Background maps are geographic features which are required to provide points of reference, for more important "foreground features". For example, if the location of bus routes is hand-drawn onto a paper map, the map acts as a background providing recognisable points of reference such as the roads, towns and rivers. In this example the bus route would be regarded as a foreground feature. The terms "background map" and "foreground feature" are usually used only in the context of computer representations of maps.
	Because the role of background maps is only to provide context for foreground features, sometimes they are stored as simple computer images. Foreground features, however, are usually represented by symbols and lines which are drawn on top of the background map each time the computer displays the map. This means that they can be selectively displayed, measured, processed and enquired upon.
BS 7567	The British Standard for the transfer of digital geographic information (see NTF).
BURISA	The British Urban and Regional Systems Association.
CAD	Computer Aided Design

CCD scanner	Charge Contact Device scanner. CCD scanners use video-camera technology to obtain images of source documents placed on a large scanning surface. These devices use powerful overhead or back-lighting to illuminate the document to be scanned. CCD scanners range from A2-A1 size.
CD-ROM	Compact Disc Read Only Memory.
Client server [system architecture]	A distributed computer system arranged such that user applications run on a "client" computer and other applications on other computers "serve" it.
Contact scanner	Uses pen plotter technology ie a small scanning head makes contact with the map surface to construct a complete image.
CSRWR	Computerised Street and Roadworks Register
Currency	The extent to which anything is kept up to date.
Database	A structured organization of records, for purposes such as automatically generating up to date reports, and answering ad-hoc queries.
Data capture	The creation of digital data from existing information sources. In the context of digital mapping, this includes digitising, direct recording by electronic survey instruments and encoding of text attributes.
Data conversion	Conversion of data into a form suitable for use in a GIS
Data transfer	Transfer of data between GIS's requires a data transfer format independent of any GIS's internal data structure.
Digital map data	The digital data required for the user to create a map on the screen or create a hard copy.
Digitizing	The conversion of analogue maps and other sources to a digital form. This may be point digitizing, where points are only recorded when pointing the cursor and pushing appropriate buttons, or stream digitizing where points are recorded automatically at pre-set intervals of either distance or time. See also scanning and vector.

Glossary

Domesday 2000
The Domesday 2000 project is part of the initiative to develop a National Land Information System (NLIS) for Britain, by the turn of the century. The project aims to create a national computerised archive of property and land data in Britain. The Domesday Research Group (DRG) was formed in November 1991 under the supervision of Professor Peter Dale to bring together research which could assist the NLIS initiative. The members of the DRG come from both GIS and property backgrounds. The project is likely to be one of the largest proposals for collecting information on land, what lies below it, upon it and above it, not only in Britain but possibly in Western Europe.

Dpi
Dots per inch. A measure of the quality or detail of scanning and printing devices.

DTP
Desk Top Publishing

Fourth Generation Language (4GL)
A very high level programming language which allows the programmer and end user to execute complex functions with only a few commands. They provide speed, flexibility and ease of use over conventional third generation languages eg COBOL and PL/1.

Fourth Generation Language tools
Examples of these range from systems enabling complex form and report definition, menu control etc from within the 4GL, down to products containing little more than commands to control the flow through the processing code and the ability to embed Query Language commands and access to forms, menus etc.

GIAG
Geographic Information Advisory Group of the LGMB (qv)

GIS
Geographic Information System. A system for capturing, storing, checking, integrating, manipulating analyzing and displaying data which are spatially referenced to the earth.

GPS
Global Positioning System. A constellation of US satellites. The satellites transmit signals which can be decoded by receivers to determine positions any in the world.

GUI	Graphical User Interface. A user interface which makes use of graphical objects, such as icons, for selection of options, and usually has a windowing capability, enabling multiple window displays on the same screen.
HAUC	Highways and Utilities Committee
IGGI	Interdepartmental Group on Geographic Information
Interoperability	The ability of two or more computer and/or information systems or their components to exchange and mutually use information.
IRDS	Information Resource Dictionary System. A metadata dictionary. A standard for data dictionaries.
LAN	Local Area Network
LGMB	Local Government Management Board
Metadata	Metadata is information about other information. The AGI is responsible for distributing the TIIWG metadata - a computerized list identifying and describing the data held by government.
MS-DOS	Micro-Soft Disc Operating System for use in Personal Computers.
Multi-media	A combination of a variety of user interfaces/communication elements such as still and moving pictures, sound, graphics and text.
Neural networks	An approach to computing whereby a computer is trained or learns to recognize patterns in data.
NDPB	Non Departmental Public Body. A body set up by Government but not directly controlled by a Government Department. A quasi autonomous Government body.
NJUG	National Joint Utilities group.
NTF	National Transfer Format. A UK Standard for the transfer of geographic data, administered by the AGI. It later evolved into a British Standard (qv BS7567).

Glossary

OPCS	Office of Population Censuses and Surveys
OSF	Open Software Foundation. A non-profit making organization aiming to produce a common operating system across hardware platforms. It supports a version of UNIX that rivals AT&T's System V.
Pen Based Computers	Pen-based computers are hand-held computer systems which are operated by the use of a pen or stylus on a flat display screen. Such systems are ideally suited for fieldwork applications such as data collection.
Real time	A real time computer system controls events as they occur.
Raster Maps	Raster maps are computerized images of maps which have been captured using scanning devices. Such maps are represented in a computers memory as a matrix of dots called pixels. Because of the way in which they are stored, they represent a very simple, economical and quick way of representing maps in computers. However, the main disadvantage is that the computer cannot distinguish between individual features of such maps.
	Raster maps are generally used as background maps (q.v.) with which the computer does not need to distinguish between individual map features. However, special software can be used to follow lines or outline areas on raster maps, allowing raster map features to be converted into a "vector" form (q.v) which is a more flexible form for computers to work with.
Scalability	The ability to host the same software environment across a wide range of computer platforms from PC to supercomputer.
Scanned maps	See Raster maps.
Secondary storage	Disc storage as opposed to a computer's main store.
SQL	Structured Query language. Originally a language for access, manipulation and querying of data in relational databases. It now has a broader role and its

	International Standard name, Database Language SQL, reflects that role.
SQL 3	The third major version of SQL likely to be ratified in 1995
SSADM	The Structured Systems Analysis and Design Method, SSADM, is CCTA's method for systems analysis and design.
Tradeable Information Initiative	An initiative run by the DTI to produce metadata, a register of geographic information sources in Government
TIIWG	Tradeable Information Initiative Working Group.
UNIX	An operating system developed by AT&T. Now adopted as an industry standard, allowing portability of applications software between hardware from different manufacturers.
Unix International	An international group of manufacturers formed in 1989 to support AT&T's Unix System V.
Vector Maps/data	Vector maps are computerized representations of maps in which individual map features are stored separately in the computer's memory. For example, the route of a river would be represented by a sequence of coordinates, which are converted into lines on the computer screen each time the map is displayed.
	Because of the way in which vector maps are stored, they offer very much more scope for processing individual features; consequently "foreground features" (q.v.) are usually stored in vector form.
	The work required to identify each individual map feature means that vector data is usually many times more expensive than raster data. Many organizations use vector data for background maps, even though they have no practical need to do this.
Wide Area Network	Telecommunications networks spanning more widely than a single site or campus.

Glossary

WIMP interfaces	Windows, Icons, Menus and Pointers. A widely used type of human computer interface.
Windows	A Graphical User Interface (GUI) built on the MS - DOS operating system that provides access to all the applications that run under it. Whilst running an application a user may summon another application which appears on the screen whilst the original is still displayed as a smaller foreground aperture , a "window".
Windows 3.x	Versions of Windows (q.v.)
Windows NT	The planned replacement for Windows 3.
X terminals	Devices dedicated to the running of the industry standard X Window System (qv).
X Windows	The X Windows system process running on a workstation, provides a display and windowing service to application processes running on the workstation or elsewhere. X Window provides the mechanism for drawing windows, it does not define a particular user interface.

Bibliography

[1] Hookam, C
GIS Problem Definition Report, 1992
National Rivers Authority

[2] GIS Markets and Opportunities. 1992
Daratech

[3] Newell, R. G.
A UK Vendor's Perspective of the GIS Market. Feb. 1992
GIS Europe

[4] Rowley, J and Rickman, D.
The Scope and Strategy for GIS Standards. 1992
Association for Geographic Information

[5] Department of the Environment 1987
Handling Geographic Information. Report on the Committee of Enquiry chaired by Lord Chorley.
HMSO ISBN 0 11 752015 2

[6] Department of the Environment 1988
Handling Geographic Information. The Government's response to the report of the Committee of Enquiry chaired by Lord Chorley.
HMSO ISBN 0 11 752080 2

[7] CCTA
Geographic Information Systems: A Buyer's Guide.
HMSO ISBN 0 11 330606 7

[8] Smith, R. (Ed.)
Dictionary of Artificial Intelligence. 1990
Collins

Index

Aerial photography 10-12, 18
AGI 1, 17, 22, 23, 28, 51, 53-56, 63, 66, 70, 72
AM/FM 25, 52
Application of emergent technology 2, 46
Applications development 55
Association for Geographic Information 17, 22, 42, 48
Atlas 34, 53
Automated mapping 5
Awareness 22-24, 26, 28, 29, 42, 43, 47, 50, 51, 53, 62-64, 69, 72

Background map   see Base map
Barriers 1, 3, 4, 42, 47, 53
Base map 8, 32
BCS   see British Computer Society
Benefits 1, 49, 51, 52, 57, 63, 69-71
British Cartographic Society 17, 27
British Computer Society 1, 17, 28
BS7567 44
BURISA 28

CAD 10
Case studies 59, 63
CCD Scanner 35, 41
CD-ROM 21, 35, 41
Central government representation 65
Chorley Report 22, 26, 38
Command and Control Systems 3, 12
Conclusions 4, 69
Cost recovery 48, 49, 57-59, 70, 71
Costs 19, 21, 31, 35, 40, 45, 49, 51, 52, 55, 57, 65
CSRWR 25, 33
Currency 6, 43, 60, 65

Data and systems 53
Data exchange 21, 44, 60
Database 2, 5, 8, 9, 21, 40
Data purchase 44, 71
Data release 45, 48, 58, 66, 71
Data suppliers 16, 18
Demographic/market survey 3, 10, 11

Desk Top Publishing 35
Digital Cartography 9
Digital Map Data 22, 38
Digitizing 19, 32, 49
Domesday 2000 29, 61
Drivers 3, 39
DTP see Desk Top Publishing

Enablers 3, 41

Foreground features 8

Geographic Information Archive 66
GIAG 23, 24, 26
Government Centre for Information Systems 1
GPS 10, 44
GUI 34, 56

HAUC 25
Hydrographic Office 22

IGGI 26, 27, 53, 65, 70, 71
Implementation
    of policies 16, 20, 24, 28, 46, 47, 55, 56, 61, 66, 67
Influencing Factors 37
Information Systems Strategy 42, 50
Inter-departmental data forum 64
IRDS 55, 56

LAN see Local Area Network
Land survey 10
LGMB 1, 17, 23, 24, 26, 63, 65
Local Area Network 41
Local Government Management Board 17, 23

Management publications 9, 25, 62
Market Testing
Metadata 53
MS-DOS 33, 34
Military Survey 21

National Address Database 19
NDPB
Neural networks 36
NJUG 1, 17, 25, 65

Index

Northern Ireland GIS Liason Committee 27

OPCS 40
Opportunities 1-4, 24, 26, 47, 50, 51, 53, 62, 64, 65, 69
Ordnance Survey 1, 5, 11, 16, 18-24, 32, 33, 47-50, 64, 66, 70
Ordnance Survey of Great Britain 16, 18
Ordnance Survey of Northern Ireland 16, 20
OS see Ordnance Survey
OS Policies 48
OSF 34

Pen-based computers 11, 36
Private sector data suppliers 22

Real-Time GIS 3, 12, 13
Representative Organizations 22
RICS 29
Royal Air Force 11
Royal Institution of Chartered Surveyors 29
Running pilots in parallel 47
Rural Wales GIS Forum 26, 64

Satellite imagery 10, 12, 20, 41
Scalability 55
Scottish GIS Forum 26, 64
Secondary Storage 35
Service providers 16
Special Interest Groups 27
SQL 36, 55
Strategy 20, 27, 42, 50, 51, 54, 55, 61, 62, 71, 72
System suppliers 16

Technology trends 2-4, 31, 33
TIIWG 26, 53
Triggers
    project-based GIS 3-4, 38

UNIX international 33-34
User and Data Forums 26
User community 17, 32

Wide Area Network 34

WIMP Interfaces 9, 41
Windows 9, 32-34

X Terminals 34